Axolotls

Axolotls care, health, diet, breeding, cages, pro's and cons and lots more included

Ben George Carre

Table of Contents

Introduction

Greetings from the enchanting realm of axolotls, where aquatic appeal and exceptional pet friendship collide. Known by many as "Mexican walking fish," axolotls are amazing aquatic animals that have won the hearts of pet lovers all over the world. We will examine the fundamentals of taking care of axolotls in this investigation, revealing the tricks to furnishing the ideal living space for these interesting reptiles.

This adventure will walk you through the nuances of taking care of these amazing pets, from comprehending their behavior and interactions to choosing the ideal Axolotl variety that matches your interests. Explore the nutritional needs that contribute to your Axolotl companion's health and learn how to maintain an ideal habitat with tank setup suggestions.

We'll cover important subjects like axolotl lifetime, aging process, and common health concerns as we make our way through the waters of axolotl ownership. This book seeks to offer insightful information on all facets of Axolotl care, regardless of your level of experience with aquatic pets.

As we investigate fascinating DIY projects to improve axolotls' underwater habitat, dispel beliefs about them, and solve riddles surrounding them, get ready for an engrossing journey. Come along for a thorough investigation of the ownership of axolotls, where every subject will enhance your knowledge and admiration of these fascinating animals.

Chapter 1

Essentials of Axolotl Care: Establishing the Ideal Habitat

Axolotl habitat creation is a complex process that includes knowing their natural habitat, putting up a suitable tank, and taking care of several aspects of their upkeep. We will go into the specifics of axolotl care essentials in this extensive book, providing in-depth insights into every facet of designing the ideal home for these unusual amphibians.

Recognizing Axolotls

Native to Mexico, axolotls are neotenic salamanders that never lose their aquatic juvenile characteristics. Understanding their natural habitat is essential to giving them the best care possible. Axolotls prefer temperatures between 60 and 68°F (15 and 20°C), and they live in calm, slowly flowing bodies of water like lakes and canals. For their welfare, these settings must be recreated.

Choosing the Correct Tank

Choosing the right tank is the first step towards constructing the ideal habitat. For an individual axolotl, a minimum tank size of 20 gallons is advised, as they need plenty of room to swim and explore. Bigger tanks are better since they provide these inquisitive animals more space to explore. Because axolotls are surprisingly good at detecting openings, make sure the tank has a tight-fitting lid to prevent escapes.

Configuring the Tank

Substance

Select a substrate that is both secure and cozy for axolotls. Sand is a well-liked choice since it has a delicate texture that is kind to their skin and presents less risk of impaction. Steer clear of gravel as it might be consumed and cause stomach problems.

Sifting

Because axolotls are sensitive to water quality, having a top-notch filtration system is essential. To keep water clean, use a filter that offers both biological and mechanical filtration. To maintain appropriate levels of ammonia and nitrate, partial water changes and routine water monitoring are crucial.

Control of Temperature

Keep the water's temperature constant within the advised range. The appropriate temperature can be set and maintained with the use of a submersible aquarium heater. Steer clear of abrupt temperature changes; axolotls can become stressed.

Luminance

Because they are light-sensitive, axolotls prefer dim lighting. To give them a sense of security, use dim lighting and hiding places like plants or décor.

Places of Hiding and Enrichment

Add plants and decorations to create an energizing space. Provide caverns or other covered areas where axolotls can hide out when they feel the need for seclusion because they enjoy hiding places. Artificial or live plants can add more hiding spots and improve the tank's attractiveness.

Maintenance of Water Quality

The health of axolotls depends critically on maintaining perfect water quality. Test the water's pH, ammonia, nitrites, and nitrates on a regular basis. To keep a clean living environment and get rid of collected garbage, do partial water changes. Purchasing a water testing kit and planning regular maintenance can help ensure your axelotls' long-term health.

Feeding Schedule

Being carnivores, axolotls mostly consume frozen or live food. Earthworms, bloodworms, brine shrimp, and other high-protein foods are frequently included in their diet. To guarantee nutritional balance, serve a varied food, and modify portion sizes according to your Axolotl's size and age. Keep an eye on how they feed, and modify the diet as necessary.

Disease prevention and health surveillance

Frequent health examinations are essential for identifying problems early on. Pay attention to your Axolotl's eating, skin tone, and overall behavior. Any alterations could point to underlying medical issues. Keep a clean tank environment to reduce the chance of infections and quarantine new additions to stop the spread of disease.

Breeding-Related Issues

If you intend to produce axelotls, be aware of the particular needs for a productive breeding program. Reduce the water's temperature gradually to simulate a cooling phase and encourage breeding. To avoid possible cannibalism, provide appropriate nesting locations and be ready to separate adults after breeding.

Typical Obstacles and Problem Solving

Care for an axolotl can provide obstacles even with your best efforts. Treat bacterial, fungal, or external parasite infections as soon as possible. If necessary, look up typical symptoms and speak with a veterinarian who has experience treating exotic animals.

In summary

A delightful endeavor, designing the ideal environment for axolotls requires careful consideration of their natural needs. Every element is vital to guaranteeing the health and well-being of these fascinating amphibians, from choosing and configuring the ideal tank to supplying enrichment, monitoring their condition, and preserving water quality. You embark on a rewarding and responsible path as a committed Axolotl fan by adopting the thorough care requirements described in this guide. You will be rewarded for your efforts when you see these extraordinary animals flourish in a setting that closely resembles their native home.

Chapter 2

Deciphering the Enigmas of Axolotl Behavior and Communication

Exploring the world of xolotl behavior and interaction reveals an enthralling mosaic of distinct characteristics and social dynamics that characterize these fascinating frogs. We shall solve the puzzles of axolotl behavior in this thorough investigation, illuminating social structures, communication, and environmental responses.

The Axolotl Social World

Despite their tendency for solitude, axolotls engage in specific social behaviors, particularly during the breeding season. Creating a setting that encourages their innate tendencies requires an understanding of these social dynamics.

Behavior in Territories

Anxolotls can be aggressive, especially in the breeding season. In particular, male axolotls may become more self-assured and defend their area from other males. In the event of a territorial conflict, having many of hiding places and visual barriers within the tank can assist lower tension.

Breeding Customs

During the breeding season, axolotls exhibit particular habits. Males may engage in courtship behaviors such as prodding and wagging their tails. If you intend to breed Axolotls, it is imperative that you become familiar with their habits. This will enable you to create the ideal environment for fruitful reproduction.

Movement-Based Communication

Axolotls use a range of movements to convey different messages to one another. Recognizing these tiny indications improves your ability to read and respond to the requirements of your Axolotl.

Wag Tail

Axolotls frequently exhibit the characteristic of wagging their tails. It performs a variety of functions, including signaling dominance, communication, and exploration. To understand the meaning of tail wagging, pay attention to the situation.

The Gills Movement

Axolotls use their gills and lungs for respiration. To assess their general health and well-being, watch how their gills move. You should look into possible causes if you see rapid or irregular gill movement, as this could be a sign of stress or respiratory problems.

Feeding Patterns

Axolotls behave differently when they are eating. They are opportunistic eaters, and you may learn a lot about their preferences and general health by watching how they eat. Make sure the food kids are eating is appropriate and diverse.

Enhancing the Environment to Promote Behavioral Stimulation

For captive axolotls to exhibit natural behaviors and avoid boredom, an enriched habitat is essential. Adding different components to their environment helps them think more clearly and engage in healthful activities.

Places of Hiding

Axolotls enjoy hiding places, for relaxation or to get away from possible threats. Add plants, caverns, and shelters to create a dynamic space that provides privacy and stimulation.

New Items and Materials

Because axolotls are inquisitive animals, you can occasionally stimulate their curiosity by introducing new things or rearrangement of their surroundings. Try varying the substrates, decorations, and textures to maintain an interesting environment for them.

Things That Float

Axolotls take pleasure in interacting with items that float on the water's surface. Including floating plants or other items helps stimulate the mind and promote inborn tendencies like examining and exploring one's environment.

Taking Care of and Getting Along with Axolotls

Even though they're not typically seen of as "hands-on" pets, some aficionados like spending time with axolotls. However, in order to reduce anxiety and protect your Axolotl's wellbeing, handling must be done carefully.

Restricted Management

Because axolotls have sensitive skin, handling them too much may cause damage to their slime coat. If handling is required, do so sparingly and with damp hands to prevent skin damage. Be kind to them at all times and watch how they react to handling.

Identifying Stress Indicators

Axolotls that are handled or exposed to strange stimuli may show signs of stress. Keep an eye out for signs such as irregular swimming, excessive gill movement, or decreased appetite. If stress is noticed, reduce the amount of time you handle your axolotl and give them time to adjust.

Good Communication via Eating

Feeding schedules help promote positive relationship. Axolotls quickly learn to correlate their owner's presence with food sources. Take advantage of feeding time to watch your axolotl and engage in non-intrusive interaction.

Lifespan and Aging of Axolotls

In order to provide the right care at various life phases, it is crucial to comprehend the lifetime and aging process of axolotls.

Lifespan

Axolotl lifespans range from 10 to 15 years on average, depending on a number of variables including nutrition, genetics, and environmental influences. Providing them with regular, excellent care extends their life expectancy.

Getting Older

Axolotls may have minor activity declines and color changes as they get older. In order to keep an eye out for any indications of age-related problems or behavioral changes, routine health examinations become more and more crucial.

In conclusion, Developing a Relationship with Axolotls Discovering the secrets of axolotl behavior and interaction exposes a world of intriguing subtleties. Connecting with axolotls is more than just owning a pet; it involves understanding their tiny motions, studying their social dynamics, and creating habitats that are enriching for them. It entails developing an awareness of their distinctive qualities and modifying care procedures to promote their wellbeing. Through embracing the subtleties of axolotl behavior, fans set out on a fulfilling journey that strengthens the relationship between these fascinating animals and the people who look after them.

Chapter 3

Selecting the Ideal Axolotl: Types and Characteristics

It's important to know the different sorts and characteristics of axolotls before bringing one into your home. Axolotls are colorful and patterned creatures with unique characteristics that add to their special appeal. We will delve into the world of Axolotl types and reveal their characteristics, maintenance needs, and things to think about while choosing the ideal aquatic friend.

Different Axolotl Types

Axolotl of the Wild Sort

The axolotl in its natural, ancestral form is known as the Wild-Type Axolotl. It usually has speckled patterns and a dark brown or greenish coloring. Many fans are drawn to the Wild-Type Axolotl because of its timeless and classic beauty, even though it is not as colorful as some other species.

Leucistic Axolotl

Leucistic axolotls have a gorgeous white or light pink coloring due to a lack of melanin. Their eye-catching look, with their bright eyes and pinkish gills, makes them a favorite among axolotl enthusiasts. Leucistics are frequently selected because of their ethereal, otherworldly appearance.

Golden Axolotl Albino

Golden Albino Axolotls are distinguished by their vivid golden or gold color. Their pigmentation reaches their gills, where it contrasts strikingly with the pinkish or reddish filaments that encase their outer gills. The Golden Albino type is highly valued because to its vibrant and striking appearance.

Axolotl melanoid

The coloration of melanoid axolotls is dark, nearly black. Melanoids have a sleek, consistent hue and aren't speckled like Wild-Type Axolotls. Their glossy, silky skin and remarkable look add to their attraction.

Axolotl copper

The distinctive color combination displayed by copper axolotls frequently consists of tones of orange, brown, and copper. Their alluring iridescence lends an air of refinement to their entire look. Axolotls with copper coloring are highly sought for due to their unique and striking appearance.

Axolotl with Green Fluorescent Protein (GFP)

Genetically engineered to express a fluorescent protein, GFP Axolotls glow in specific light conditions. This variety has the captivating bonus of bioluminescence along with a range of base colors, such as Wild-Type, Leucistic, and others.

Axolotl Chimera

Due to a unique and fascinating genetic abnormality, chimera axolotls have bodies that combine many color patterns. These exceptional individuals resemble mosaics, which makes each one a genuinely unique specimen.

Characteristics and Things to Think About

Dimensions

Selecting the appropriate axelotl size for your tank and living area is essential. Axolotls come in a variety of sizes. Though most grow to be between 9 and 12 inches long, some could get larger. When choosing the appropriate Axolotl for your home, take into account the tank's capacity as well as the animal's future growth.

Age

There are several age categories of axolotls, ranging from juveniles to adults. Because they are smaller than adult axolotls, juveniles can need specialized care. When choosing the age of your Axolotl, take your tastes and experience level into account.

Sex

It can be difficult to determine the gender of an axolotl, particularly in their early years. On the other hand, males may have more streamlined features and more noticeable cloacal glands than females during the breeding season. In your selecting process, gender could have a less role if breeding is not an issue.

Well-being

Choosing a healthy axolotl is essential to having a happy and successful pet ownership experience. Seek out individuals with transparent skin, undamaged gill filaments, and vigorous swimming habits. Steer clear of Axolotls that appear injured, ill, or lethargic.

Harmony

Think about how well-suited each Axolotl is if you intend to retain more than one. Even though axolotls are solitary animals in general, there are variations in their temperaments. Providing enough hiding places and introducing them gradually can help prevent confrontations.

Considerations for Selecting an Axolotl

Education and Research

Take the time to learn about the particular upkeep needs of the Axolotl variety you're interested in before making your purchase. Regarding temperature, tank capacity, and nutrition, different types could require slightly different conditions.

Source and Reputation of Breeders

Choosing a trustworthy breeder or supplier is essential to getting a healthy, well-maintained xolotl. Seek out breeders who have a track record of satisfied customers, moral breeding methods, and a dedication to the welfare of their pets.

Tank Configuration

Before bringing an axolotl home, make sure your tank is properly set up and cycled. In terms of water temperature and tank conditions, different types may have slightly different preferences, so adjust your setup to suit the unique requirements of the variety you've chosen.

Legal Aspects to Take into Account

Verify local laws regarding Axolotl ownership, as certain regions can have limitations or call for permits. Before you purchase an Axolotl, make sure you are in compliance with all applicable laws.

In summary

Selecting the ideal Axolotl requires careful evaluation of a number of variables, including compatibility and practical care requirements as well as the aesthetic appeal of various types. Enthusiasts can make decisions that promote the health and happiness of these fascinating aquatic companions by being aware of the peculiarities and distinctive qualities of each kind. For individuals prepared to start a fulfilling adventure of Axolotl ownership, the world of Axolotl variations offers a broad assortment of options, whether they are drawn to the ethereal beauty of Leucistics, the colorful attraction of Golden Albinos, or the sleek elegance of Melanoids.

Chapter 4

Axolotl Nutrition: A Comprehensive Guide for Maximum Well-being

A vital part of your Axolotl's care that greatly enhances their general health and wellbeing is feeding them. Since axolotls are carnivorous amphibians, feeding them a food rich in nutrients and well-balanced is essential for their immune system, growth, and general well-being. We will look at the dietary requirements of axolotls, appropriate food choices, feeding methods, and advice on how to keep your pet healthy with a well-planned diet in this extensive guide.

Recognizing Axolotls' Nutritious Requirements

Since axolotls are carnivores, their main source of food is animal stuff. tiny water invertebrates, worms, insects, and even tiny fish make up its food in the wild. To suit their nutritional needs, it is imperative to replicate this diet in captivity.

Important Elements

Protein: To ensure healthy growth and maintenance of internal systems, axolotls need a diet high in high-quality protein. Live or frozen foods like brine shrimp, earthworms, bloodworms, and small fish can be sources of protein.

Fat: Despite being relatively low-fat creatures, axolotls yet require vital fats in order to function. Usually, feeding them a range of foods high in protein will supply enough fat.

Minerals and vitamins: Eating a varied diet helps provide the range of nutrients needed for overall health. Specifically, calcium is essential for healthy bones and can be taken as a supplement if necessary.

Appropriate Meals for Axolotls

Living Foods

Earthworms: Rich in protein and easily digestible, earthworms are a mainstay in the diet of axolotls.

Bloodworms: Axolotls easily swallow these tiny, scarlet larvae, which are rich in protein.

Brine Shrimp: Frequently found frozen, brine shrimp are a great source of protein and can be thawed before to feeding.

Blackworms: Axolotls can eat these slender, aquatic worms as a wholesome live food source.

Cold Foods

Frozen Bloodworms: A practical alternative to live bloodworms, the frozen kind is also available.

Frozen Brine Shrimp: Frozen brine shrimp are a useful addition to their diet because they are easily stored and thawed.

Frozen Fish Fillets: Providing tiny, thawed fish fillets adds diversity and extra nutrients, and is ideal for larger Axolotls.

Pelts

Sinking Pellets: Axolotls are the target audience for certain premium sinking pellets. To avoid choking, make sure the pellets are the right size for your axolotl.

Pellets Particular to Axolotls: Axolotl-specific specialty pellets often include a well-balanced blend of nutrients. Include these in a diversified diet.

Controlling Portion and Frequency

Axolotls' feeding schedule varies according to their size and age. While immature axolotls could need more frequent feedings, adult axolotls can often be fed two to three times each week. To prevent overfeeding, pay attention to their physical state and modify the frequency and portion sizes accordingly.

Feeding Methods and Things to Think About

Manual Feeding

A gratifying method of bonding with your axolotl and keeping an eye on their eating patterns is to hand-feed them. Place food in front of them immediately using your fingers or feeding utensils. Axolotls might take a while to find and eat their food, so be patient.

Size and Preparation of Prey

Based on the size of your axolotl's mouth, change the size of the prey objects. Smaller animals could find it difficult to handle larger prey, which could result in choking dangers. Provide adequately sized prey items to larger axolotls to promote their natural hunting behavior.

Diverse Foods

To provide a wide range of nutrients, a diversified diet is necessary. Alternate your Axolotl's diet between frozen and live meals to make sure it gets a well-rounded diet. This makes mealtimes engaging for them while also promoting their health.

Supplementing with Calcium

Calcium is necessary for axolotl bone formation and maintenance. Although most of the time a diversified diet will supply enough calcium, you can think about dusting prey items with an occasional calcium supplement, particularly for growing juveniles and gravid females.

Particulars Regarding Axolotl Diets

Young Axolotls

Due to their essential growing phase, juvenile axolotls have different nutritional needs than adult axolotls. Feed them more often with smaller prey items to aid in their quick growth. Keep a close eye on their progress and modify the diet as necessary.

Grasping Ladies

A slightly modified diet may be necessary for female axolotls to support their reproductive demands as they get ready to produce eggs. Provide foods that are high in nutrients and think about having meals more frequently to meet the increased energy needs during this time.

Keeping an eye on and modifying diet

Keep a close eye on the general health, weight, and behavior of your Axolotl. Adapt their diet in response to any noticeable alterations. If an axolotl starts to gain weight, cut back on feeding schedule and be careful to watch portion sizes.

Concluding Remarks: Promoting Maximum Health with Diet

Feeding your axolotl is an essential part of being a good pet owner, not just a daily chore. You enhance their general health and lifespan by being aware of their nutritional demands, providing a varied and well-balanced diet, and attending to their specific needs. The link between caregiver and aquatic companion strengthens as you begin the process of providing sustenance for your Axolotl. This tie is based on understanding, compassion, and the delight of seeing these unusual amphibians flourish in captivity.

Chapter 5

Setting Up an Axolotl Tank: Advice for a Content and Healthy Pet

Establishing the perfect tank environment is essential to your Axolotl's health and pleasure. Since axolotls are aquatic amphibians with particular environmental needs, it's critical to provide them with a home that closely resembles their natural surroundings. We'll go over all the essential components of an Axolotl tank setup in this extensive guide, including substrate, lighting, decorations, and tank size and filtration.

Dimensions & Size of the Tank

Choosing the right tank size is one of the most important things you can do to ensure the health of your axolotl. Given that axolotls are known for living in water, a large tank enables them to swim, explore, and behave like they would in the wild.

Minimum Tank Size: Twenty gallons is the minimum recommended tank size for an adult axelotl. On the other hand, larger tanks—thirty gallons or more—are better since they offer more room for swimming and a more richer environment.

Taking Into Account the Need for a Larger Tank: If you intend to keep more than one axolotl, make sure the tank is large enough. It's a good idea to add 10 liters for every extra axolotl.

Tank Dimensions: To fit the extended body of the axolotl, choose a tank that is at least 36 inches long. Longer tanks are better for their general health and offer greater swimming area.

System of Filtration

Your Axolotl's health depends on having clean, well-filtered water. Given that axolotls are sensitive to water quality, a top-notch filtration system helps eliminate waste and maintains a steady aquatic habitat.

Choosing a Filter: Pick a filter that offers biological and mechanical filtering. For Axolotl aquariums, canister or sponge filters are common options. Because axolotls prefer water that moves slowly, make sure the filter's flow rate is not too vigorous.

Frequent Maintenance: To avoid debris accumulation and preserve the best possible water quality, clean and maintain the filter on a regular basis. Keep an eye on water quality measures like ammonia, nitrite, and nitrate levels. When necessary, do partial water changes.

Selection of Substances

In addition to offering the axolotl a comfy area to land on, the substrate in your tank harbors helpful bacteria that aid in biological filtration.

Sand Substrate: Axolotl aquariums often use sand as their substrate. It lessens the chance of harm because it is soft and delicate on their skin. Axolotls may also consume substrate while hunting, however sand is less dangerous than gravel.

Steer clear of gravel: If swallowed, gravel can induce impaction and present a choking threat. In order to protect your pets, it is advised that you stay away from using gravel substrates in your Axolotl tanks.

Cleaning Ease: Sand is a relatively low maintenance material. When changing the water, gently suction the substrate to get rid of detritus without upsetting the axolotl.

Temperature and Heating of the Water

It's critical to maintain the right water temperature for axolotl health. Axolotls prefer colder water, therefore it's best to keep the temperature as constant as possible.

The ideal temperature range for water is between 60 and 68°F (15 and 20°C). To get and keep the right temperature, use a submersible aquarium heater. Axolotls should not be subjected to abrupt temperature changes.

Thermoregulation: Because axolotls are ectothermic, their environment affects how hot their bodies get. Their metabolism and general health are supported when the water is kept at a consistent and appropriate temperature.

Luminance

Even though axolotls don't need special illumination, having enough lighting in the tank is still functional and improves the overall appearance.

Low-Intensity Lighting: To replicate the axolotl's native environment, use low-intensity lighting, as they are sensitive to intense light. Soft LED lighting or daylighting in general are good choices.

Lighting Schedule: To create a day-night cycle, have a regular lighting schedule. This gives your axolotl a feeling of routine and helps control their circadian rhythm.

Decorations and Enhancement
Establishing a well-lit space with appropriate accent pieces and hiding places is crucial for your Axolotl's welfare. Stress is decreased and natural behaviors are encouraged by enrichment.

Hiding Spots: To provide your axolotl a place to hide, include caverns, shelters, and vegetation. Live or artificial plants, ceramic hides, and PVC pipes can provide both security and enrichment.

Axolotls take pleasure in interacting with things that float on the water's surface. Including floating plants or other items helps stimulate the mind and promote inborn tendencies like examining and exploring one's environment.

Aquatic Decorations: To avoid injuries, use decorations that are rounded and smooth. Because axolotls may scratch surfaces, it's important to make sure the decorations are safe.

Maintenance of Water Quality

Keeping the water at its ideal quality is essential for your axolotl's wellbeing. Effective water quality management primarily consists on routine testing and water changes.

Water Testing: To keep an eye on ammonia, nitrite, nitrate, pH, and other pertinent factors, use a dependable water testing kit. Make sure the water is suitable for axolotls by testing it on a regular basis.

Partial Water Changes: To eliminate collected debris and restore vital minerals, do partial water changes on a regular basis. The size of the tank, the effectiveness of the filtration, and the quantity of axolotls all affect how frequently the water is changed.

Positioning and Stability of Tanks
The placement of your axolotl tank has an impact on their general health. Assure environmental stability and reduce disruptions to a minimum.

Steer Clear of Direct Sunlight: To avoid algae growth and temperature swings, keep the tank out of direct sunlight.

Stable Environment: Steer clear of locations for the tank where there are a lot of loud noises or vibrations. Because axolotls are sensitive to disturbances, they feel more at ease in a steady, quiet environment.

New Additions and Quarantine

Keep any new additions to your axolotl tank quarantined to avoid introducing any potential parasites or diseases. You may keep an eye on the health and behavior of newly acquired Axolotls by observing them in a separate tank prior to integrating them into the main tank.

In conclusion: Establishing an Axolotl Haven

Creating a sanctuary that meets your axelotl's specific requirements and enhances both their physical and emotional wellbeing is more important when designing a tank setup for them than just giving them a somewhere to live. You may start your journey of responsible Axolotl ownership by putting these tips for a happy and healthy pet into practice. This will help you develop a relationship with these fascinating aquatic companions that is based on compassion, understanding, and the satisfaction of seeing them thrive in their specially designed aquatic environment.

Chapter 6

Recognizing the Lifespan of Axolotls and Aging Well

In order to provide the right care and guarantee the wellbeing of these fascinating aquatic animals, it is imperative to comprehend the lifetime and aging process of axolotls. Because of their special neotenic traits, axolotls have a unique life cycle that calls for particular attention at certain points. This thorough guide will cover the average lifespan of axolotls, variables that affect their longevity, aging indicators, and advice on how to help these amazing amphibians age gracefully.

Duration of Axolotl Life

In ideal circumstances, axolotls can live for ten to fifteen years on average. However, a number of factors, including as nutrition, environment, genetics, and the caliber of care they receive, affect how long they live.

Caretakers can improve the general health and longevity of their Axolotl friends by being aware of these aspects.

Molecular Biology

An axolotl's possible lifespan is mostly determined by genetics. The general health and lifespan of an individual can be influenced by both the quality of their breeding and their genetic origin. Acquiring Axolotls from respectable breeders who emphasize ethical breeding methods raises the possibility of obtaining healthy animals with extended potential life spans.

Nutrition and Diet

Axolotls depend on healthy diet to stay healthy and live long lives. Overall wellbeing is supported by a nutritionally adequate, well-balanced diet. Offering a range of live and frozen feeds, including brine shrimp, bloodworms, and earthworms, guarantees that Axolotls get the nutrition they need for healthy growth and immune system operation.

The surroundings

The environment in the axolotl tank has a direct effect on how long they live. It is crucial to keep the temperature, pH, and quality of the water constant. A healthy aquatic habitat is facilitated by regular water testing, adequate filtration, and periodic maintenance, which lower stress and any health problems that could shorten their lives.

Medical Care

Axolotl care requires regular health checks and quick reaction to any indications of illness or distress. Living a longer and healthier life can be attributed to identifying and treating health issues at an early age. If you have any worries regarding the health of an axelotl, speak with a veterinarian who has experience with exotic pets.

Axolotls' Aging Signs

As they get older, axolotls show subtle indications of aging, and their caregivers need to be alert to spot these changes. Comprehending the ageing process of Axolotls facilitates the provision of suitable care and the modification of the surroundings to suit their requirements.

Variations in Hue

Axolotls may have minor color changes as they get older. The markings of juvenile Axolotls are frequently bright and distinct, but as they age, their colors may fade or become duller. This typical, natural shift in pigmentation is not always a sign of a medical problem.

Diminished Involvement

Axolotls that are getting older could be less active than their younger, more active counterparts. Even while axolotls are typically sedentary creatures, a noticeable decrease in mobility or lethargic behavior may indicate possible health issues. Tracking variations in activity levels can aid in the early detection of problems.

Diminished Hungry Feeling

Axolotls that are getting older may become less hungry. There are several possible causes for this, such as modifications in metabolism or problems with the teeth. Age-related variations in appetite can be addressed with diet adjustments, by providing smaller and softer prey items, and by keeping an eye on feeding behavior.

Alterations in Gills

Axolotls' gills are necessary for breathing, and as they age, they may exhibit symptoms of diminished gill function. Monitoring gill movement and keeping an eye out for abnormalities, including increased breathing effort, can help detect respiratory problems that could develop as people age.

Growth Valley

During their juvenile years, axolotls develop significantly; however, as they become older, this growth slows down and eventually reaches a plateau. When an axolotl reaches adulthood, its size usually stabilizes, thus any abrupt changes in growth patterns may require veterinarian intervention.

Skin Type

Axolotls that are becoming older may have altered skin textures. While the skin of younger Axolotls is usually elastic and smooth, wrinkles or folds may appear on elderly Axolotls. Skin health can be supported by drinking enough water and making sure the water is acceptable.

Encouraging Aplomb in Aging Axolotls

Axolotls can age gracefully if their caregivers take proactive steps to ensure that they have comfortable and happy lifestyles for the duration of their lives.

Sustain Ideal Water Conditions

Ageing axolotls need stable and ideal water conditions to stay healthy. Maintaining regular cleanliness, testing water parameters, and keeping an eye on filtration effectiveness will help you establish a stress-free and health-promoting environment.

Adjust Food to Needs of Aging

Axolotls' dietary needs may vary as they get older. To account for any variations in metabolism, appetite, and digestive capacity, modify their diet. To support their nutritional health, give them softer prey items and add key nutrients to their diet.

Establish a Richer Environment

The Axolotl's quality of life can be improved by providing appropriate decorations, hiding places, and a variety of substrate for their habitat. More areas to rest and easy access to the water's surface for breathing may be beneficial for elderly people.

Frequent Health Examinations
Examine the health of elderly Axolotls on a regular basis to see if there are any behavioral, physical, or eating changes. Early health issue identification enables timely response and suitable veterinarian care.

Things to Think About When Breeding Axolotls
If you are thinking of breeding axolotls, keep in mind the possible strain and energy requirements it may put on people, especially the elderly. It is important to protect the health and welfare of both the male and female axolotls during the physically demanding process of breeding.

Slow Shifts in the Environment

Make adjustments to the Axolotl tank configuration gradually. Stress can be brought on by abrupt changes in routine or surroundings, particularly in older people. Transitions should be gradual to allow for adjusting time.

Frequent Communication

Even though axolotls aren't often very talkative pets, constant observation and gentle interaction can help keep the relationship between the caregiver and the animal intact. This makes it possible for caregivers to identify any behavioral changes that might point to age-related problems.

In conclusion, Taking Care of Aging Axolotls

Axolotl lifetime and aging processes are important concepts to grasp in order to responsibly own a pet. Caretakers can help these special aquatic animals age gracefully by identifying the indications of aging, modifying care procedures to suit their evolving needs, and creating a secure and stimulating environment. As caretakers age, their link with rescued Axolotls grows, and their shared dedication to their welfare guarantees that they lead comfortable and happy lives for the duration of their incarceration.

Chapter 7

Axolotl Health Check: Typical Problems and Avoidance

Regularly checking your axolotl's health is an essential part of being a conscientious pet owner. Caretakers can detect any health problems early on and provide appropriate treatment by being proactive and perceptive. We will go over typical health problems that could affect axolotls, symptoms of illness, precautions to take, and advice on how to keep these fascinating aquatic animals in good condition in this extensive guide.

Typical Health Problems with Axolotls

1. Skin Wounds and Lesions:

- Causes: Rough handling, aggressive tank mates, and sharp items can all result in cuts, abrasions, or other injuries.

- Warning signs include redness, visible sores, and textural changes.
- Preventive measures include using delicate tank decorations, handling axolotls carefully, and making sure tankmates get along.

2. Fungal Diseases:
- Causes: Injuries, stress, and poor water quality can all result in fungal infections.
- Signs: Growth on the skin, fins, or gills that resembles cotton.
- Prevention: Keep the water clean, keep an eye on its characteristics, and create a stress-free atmosphere.

3. Bacterial Diseases:
- Causes: Wounds can become infected with bacteria, especially in stressed or injured axolotls.
- Redness, edema, or open sores are warning signs.

- Precautions include keeping the tank clean, avoiding crowding, and giving the animal the right food.

4. Infestations of Parasites:

- Causes: Axolotls can contract parasites such as flukes or protozoa.
- Warning signs include weight loss, unusual behavior, or obvious parasites.
- Prevention: Feed a varied and nutrient-rich food, keep up excellent hygiene, and quarantine new additions.

5. Impact:

- Causes: Eating big foods or substrates that can clog pipes.
- Symptoms include constipation, sluggishness, or uneven floating.

- Precaution: Provide prey that is the right size, use sand substrate, and keep an eye on eating habits.

6. MBD, or metabolic bone disease:
- Causes include inadequate food, low calcium or vitamin D levels, and bad water quality.
- Indications: malformed appendages, dysphagia, or fatigue.
- Prevention: Provide a diet that is well-balanced, give calcium supplements as needed, and keep the right water parameters.

7. Digestive Disorders:
- Causes: An improper diet, over feeding, or abrupt dietary adjustments.
- Indications include bloating, constipation, or altered fecal production.

- Prevention involves monitoring bowel motions, avoiding overfeeding, and providing a varied and balanced diet.

8. Issues with the Respiratory System:
- Causes: Respiratory diseases, low oxygen levels, and poor water quality.
- Symptoms include coughing, gill discolouration, and labored breathing.
- Prevention: Keep an eye on water parameters, maintain oxygenation, and make sure sufficient filtration is in place.

Axolotl Illness Symptoms
- It is essential to recognize the symptoms of disease in axolotls in order to diagnose them early and take appropriate action. Frequent observation and expertise with typical behavior aid caregivers

in recognizing abnormalities that might point to medical problems.

Modifications in Conduct:

- Normal Behavior: Axolotls move slowly and calmly, and they are primarily nocturnal creatures.
- Lethargy, strange swimming patterns, or hiding a lot are symptoms of illness.

Diminished Appetite:

- Normal Behavior: Axolotls in good health usually eat a lot.
- Refusal to eat, a decline in appetite, or spitting up food items are indications of illness.

Strange Things in Gills:

- Normal Behavior: Smooth and regular movement of the gills.

- Changes in gill color, increased gill motility, or obvious damage are indicators of illness.

Changes in the Skin and Body:

- Normal behavior includes elastic, smooth skin and a constant bodily form.
- Lesions, color changes, bloating, or deformities are indicators of illness.

Breathing Problems:

- Regular Behavior: Consistent, even breathing.
- Coughing, gasping, or difficult breathing are symptoms of illness.

Variations in the Fecal Output:

- Normal Behavior: Consistent bowel motions.
- Changes in the color, consistency, or frequency of feces are indicators of illness.

Floating Problems:

- Normative Behavior: Consistent and regulated buoyancy.
- Symptoms of illness include trouble maintaining a specific depth and uneven floating.

Alterations in Behavior During Feeding:

- Normal Behavior: Engaging and getting excited when eating.
- Symptoms of illness include a lack of appetite and trouble catching prey.

Preventive Actions for the Health of Axolotls

Preserve Clean Water:

- Test and keep an eye on water parameters including pH, ammonia, nitrite, and nitrate on a regular basis.
- To keep waste out of the water and preserve ideal water quality, do regular water changes.

Offer a Well-Balanced Diet:

- To guarantee a balanced and nutrient-rich diet, provide a range of fresh and frozen meals.
- Steer clear of overfeeding and modify portion amounts according to the size and age of the axolotl.

Prevent Crowding:

- To lower stress, make sure there is enough room and don't crowd people.
- To stop the spread of illness, quarantine new additions before adding them to the main tank.

Employ the Proper Substrate:

- Select a substrate made of soft sand to avoid injuries and impaction.
- Don't use gravel—it can lead to stomach problems and be a choking hazard.

Preserve Ideal Tank Conditions:

- Use the proper filtering devices to guarantee that the water is pure and has enough oxygen.
- Reduce stress by creating hiding places and improving the surroundings.

Use safe handling techniques:

- Take caution when handling axolotls to avoid stress or injury.
- Before handling them, moisten your hands to prevent harming their sensitive skin.

Frequent Health Examinations:

- Keep a close eye on your Axolotl to notice any changes in activity, look, or behavior.
- Conduct regular health examinations, looking at the skin, gills, and general state of the body.

See a Veterinarian for Advice:

- Consult a veterinarian with experience treating exotic animals for expert assistance.
- Make time for routine examinations to keep an eye on your Axolotl's general health.

New Items in Quarantine:

- Before bringing fresh axelotls into the main tank, place them in quarantine.
- To stop the spread of any dangerous diseases, keep an eye on their health and behavior in a different tank.

Enrich the Environment Appropriately:

- Incorporate plants, decorations, and hiding places to make the space lively.
- Occasionally rotate and swap out the elements to keep things interesting.

Initial Care for Axolotls

- Caretakers can administer first aid immediately in the event of minor accidents or health issues to promote recovery and offer temporary relief:

Salt Rooms:
- Use aquarium salt (no iodine) to create a gentle salt bath.
- For a brief period of time, submerge the axolotl in the salt bath to help with wound healing and infection prevention.

Separation:
- If an illness appears, place the afflicted axolotl in a different quarantine tank.
- This enables targeted care and stops any illness spread to other tank mates.

Temperature Modification:

- Lower the water's temperature to approximately 60–65°F (15–18°C), which is the acceptable range.
- Lower temperatures can promote the healing process and make people feel less stressed.

Soft Meals:

- Provide smaller portions of food or finely cut game to meet for feeding issues.
- This guarantees that the axolotl will have access to vital nutrients while it heals.

5. Observing:

- Pay special attention to the behavior, hunger, and general health of the affected axolotl.
- Keep an eye out for any symptom improvement or deterioration.

6. Speak with a Veterinarian:

See a veterinarian with experience caring for frogs for professional help if the health problem continues or gets worse.

A veterinarian is qualified to make an accurate diagnosis and suggest appropriate courses of action.

In summary, proactive maintenance of healthy axelotls Axolotl health and well-being are mostly dependent on caregivers, who do this by conducting routine health examinations, taking preventative measures, and acting quickly when necessary. Caretakers can prolong the life and well-being of these unusual aquatic friends by keeping an eye out for symptoms of illness, giving a healthy diet, and setting up a suitable environment. As proactive care methods become habitual, the relationship between the caretaker and the axolotl strengthens. It is based on mutual understanding, accountability, and the delight of seeing these amazing amphibians flourish in captivity.

Chapter 8

Raising Axolotls: An Intriguing Adventure in Procreation

Starting an Axolotl breeding operation is an exciting activity that enables caregivers to see the complex reproductive process in these unusual aquatic frogs. A fascinating opportunity to explore the mysteries of aquatic life, rearing axolotls offers everything from learning the reproductive anatomy and behavior to setting up ideal breeding conditions and tending to the eggs and larvae. We will examine every facet of axolotl breeding in this in-depth guide, from setting up the breeding habitat to rearing the next generation of these amazing animals.

Anatomy of Reproduction and Sexual Dimorphism

Understanding axolotl reproductive architecture and sexual dimorphism symptoms is crucial before beginning the breeding process.

1. Dimorphism in Sexuality:

- Male and female axolotls are essentially the same from the outside, especially when they are not reproducing.
- The cloaca is a distinguishing feature shared by both sexes, though it may seem significantly larger and more bloated in males during the breeding season.

2. Cloaca:

- The reproductive, digestive, and urine systems all exit through the cloaca.
- The cloaca is essential for the transmission of eggs and sperm during reproduction.

3. Behavior during Courtship and Mating:

- During courtship, the male will prod the female's cloaca with his nose.

- After courtship, the male deposits a spermatophore, or packet of sperm, which the female takes up with her cloaca in order to fertilize internally.

Establishing the Optimal Breeding Conditions

The first step in successful breeding is to create an environment that is similar to what occurs naturally during reproduction. The perfect breeding habitat for axolotls is influenced by several important elements, including:

Tank Configuration:

- Pick a roomy tank that is ideally bigger than the standard enclosure to allow plenty of room for mating and egg laying.

- To help the breeding couple feel less stressed, make sure the tank includes the right amount of decorations and hiding places.

Warmth:

- To replicate the start of spring and mark the beginning of breeding season, slightly warm the water.
- Raising the temperature to approximately 18–20°C (64–68°F) can promote breeding behavior.

Photoperiod:

- Modify the illumination to replicate variations in the length of a typical day.
- Increase the photoperiod gradually until it reaches about 14 hours of light, simulating longer days and indicating the arrival of spring.

Water Purity:

- In order to lessen stress and promote the health of the mating couple, keep the water in perfect condition.
- It is essential to regularly replace the water and monitor the water's pH, ammonia, nitrite, and nitrate levels.

Harmony:

- Based on their size, genetic variety, and general health, choose breeding couples that are compatible.
- To reduce stress during the breeding phase, avoid adding more tank mates.

Overview and Courtship:

- Put the axolotls—male and female—into the breeding tank.

- Keep an eye out for the male's spermatophores being deposited and his prodding and circling actions during courtship.

Watching the female clit:

- Keep an eye out for swelling in the female's cloaca, which indicates a successful courting and the existence of eggs.
- Internal fertilization occurs in the eggs after the female has picked up the spermatophores.

Laying of Eggs and Taking Care of Them

The female axolotl begins the egg depositing phase following a successful mating. To guarantee a successful hatching, the eggs must be handled and monitored carefully during this crucial time.

Site of Egg Laying:

- Make sure the structures or surfaces are appropriate for the female to lay her eggs on.
- Egg-laying sites can be artificial plants, flat stones, or PVC pipes.

Gathering the Ovum:
- After the eggs are laid, gently gather them up and move them to a different container with the right water conditions.
- Use a soft siphon or other similar instrument so as not to break the fragile eggs.

Configuration for Incubation:
- Assemble an incubation container with clean water that satisfies the breeding tank's specifications and temperature.
- For water circulation, use a sponge filter or a mild airstone.

Stopping Fungus:

- Methylene blue can be added to the incubation water to stop fungi from growing on the eggs.
- To prevent fungus, keep the eggs in a low-light setting.

Tracking the Development of Eggs:

- Keep an eye out for any signs of growth in the eggs.
- Unfertilized eggs may become opaque, whereas fertilized eggs will show a distinct, visible embryo.

Care of Larval Hatching:

- The eggs will hatch and release tiny Axolotl larvae after around 10–14 days.
- Give the larvae appropriate diet, like freshly hatched brine shrimp or commercial axolotl chow that has been finely crushed.

Setup of the Larvae Tank:

- Provide the larvae with a separate tank with the right water conditions.
- Make sure the juvenile axolotls have hiding places and gentle filtering in the tank.

Transitioning Gradually to an Adult Diet:

- Gradually provide a diet fit for juvenile and subsequently adult axolotls as the larvae get bigger.
- Keep an eye on development and growth to provide adequate nourishment.

Obstacles & Things to Think About When Breeding Axolotls

Although raising axelotls can be a fulfilling endeavor, there are certain difficulties and things to keep in mind that caregivers need to be aware of:

Carnivory:

- Even as adults, axolotls have been known to engage in cannibalistic behavior, particularly with regard to eggs and small larvae.
- To reduce the possibility of cannibalism, give eggs and young axolotls lots of places to hide.

Genetic Points to Consider:

- Genetic variety must be taken into account and inbreeding must be avoided in responsible breeding.
- Don't breed near relatives and keep track of the breeding couples.

Monitoring Health:

- Keep an eye on the developing eggs' and the breeding pair's health.
- If you notice any indications of stress or illness, be ready to step in and help.

Variations in Temperature:

- The health of eggs and larvae as well as the success of reproduction might be affected by abrupt temperature fluctuations.
- To aid in the breeding process, keep the temperature steady.

Accessibility of Appropriate Food:

- Make sure the larvae have a consistent supply of wholesome food.
- Make sure there will be live or frozen food available for the developing axolotls in advance.

Duration & Dedication:

- Axolotl breeding takes patience, dedication, and close observation.
- Be ready to shoulder the obligations that come with taking care of the breeding pair and their progeny.

Taking Into Account Several Clutches:

- If the female produces more than one clutch of eggs, think about giving her enough time to recover in between breeding cycles.

Final Thoughts: Exploring Axolotl Procreation

Axolotl breeding provides caregivers with a unique opportunity to see the life cycle of these intriguing aquatic creatures, making it an enthralling voyage into the realm of reproduction. The process takes meticulous planning, dedication, and monitoring, from comprehending the anatomy and behavior of the reproductive system to establishing the ideal breeding habitat and tending to the eggs and larvae. Caretakers can help save axolotls and get a greater appreciation for the beauties of aquatic life by accepting the challenges and thinking about the health of the breeding pair and their young. Axolotl breeding offers the opportunity to actively contribute to the preservation of these amazing amphibians in addition to being an educational experience.

Chapter 9

Dispelling Axolotl Myths: Differentiating Real from Fiction

Because of their unusual look and intriguing habits, axolotls are frequently the subject of various myths and misunderstandings. Navigating the deluge of data is essential for conscientious Axolotl ownership. We'll dispel widespread misconceptions about axolotls in this extensive guide, offering factual information to assist caregivers in making decisions on the upkeep and welfare of these fascinating aquatic animals.

Myth 1: Fish Are Axolotls

Disproving:

An axolotl is an amphibian, not a fish. In particular, they are closely related to salamanders and are members of the order Caudata. Axolotls are vertebrates with limbs, lungs, and gills, unlike fish. While they go through metamorphosis in the nature, they frequently remain in their aquatic juvenile form in captivity for the whole of their lives, a trait called neoteny.

Myth 2: Any Body Part Can Be Regenerated by Axolotls

Disproving:

Although axolotls are well known for their remarkable regenerative powers, it is untrue to say that they can regenerate any part of the body. Axolotls have the ability to regenerate parts of their brains, hearts, spinal cords, and limbs. Nevertheless, their capacity for regeneration is limited. They are unable to regenerate intricate organs such as the eye or specific parts of the head.

Myth 3: Axolotls Do Not Experience Transformation

Disproving:

Axolotls have the ability to change into different species, even though they are neotenic and can sometimes retain their aquatic juvenile characteristics. Metamorphosis is a normal aspect of life for creatures in the wild, brought on by environmental elements including temperature, water quality, and population density. When kept in captivity, axolotls can maintain their neotenic state if given the right environmental circumstances.

Myth 4: You Can Keep Axolotls in Tiny Bowls
Disproving:
Keepers of axolotls should avoid keeping them in small bowls as they need large, well-maintained homes. Axolotls require a sufficiently large tank, against popular belief, in order to facilitate their swimming and exploring. A bigger tank eases stress, promotes general wellness, and helps keep water conditions steady. A single adult axolotl is thought to require a minimum of a 20-gallon tank.

Myth 5: Filtration Is Not Needed for Axolotls

Disproving:

Because axolotls are sensitive to water quality, keeping an environment free of contaminants requires adequate filtration. It's a fallacy that axolotls don't require filtering. Filtration aids in the removal of pollutants, trash, and dangerous materials from water, resulting in a stable and clean environment. It is essential to use the right filter in order to promote axolotl health and well-being.

Myth 6: Warm Water Is Perfect for Axolotl Survival

Disproving:

Being cold-water amphibians, axolotls prefer colder climates for survival. It is untrue to say that axolotls can live in warm water. It is vital to keep them at a healthy temperature range of 60–68°F (15–20°C). Increased temperatures can stress axelotls, have an impact on their metabolism, and raise their likelihood of developing health problems. It's critical to provide them a consistent, cool atmosphere for their general health.

Myth 7: Axolotls Don't Require Decorations or Hideouts
Disproving:
A nicely decorated tank with hiding places and appropriate décor is beneficial to axolotls. It's a fallacy that axelotls don't require decorations or hiding places. Adding plants, caverns, and hiding places to their surroundings improves their quality of life while lowering stress levels. Because axolotls may withdraw to hiding places in times of stress or danger, these characteristics are critical to their mental health.

Myth 8: Water Changes Are Not Needed by Axolotls
Disproving:
To keep the water quality in an axolotl tank at its best, regular water changes are essential. It is untrue to say that axolotls don't require water changes. Poor water

quality can result from garbage, uneaten food, and other contaminants building up over time. Frequent water changes promote the health of axolotls by removing these contaminants and replenishing vital minerals.

Myth 9: Diverse Diets Are Not Necessary for Axolotls
Disproving:
Diets that are varied and well-balanced are beneficial to axelotls. It is a fallacy that axolotls don't need to eat a variety of foods. Providing a variety of live and frozen feeds, like brine shrimp, bloodworms, and earthworms, guarantees that they get the nutrition they need. Their ability to reproduce, grow, and maintain general health are all enhanced by a diversified diet.

Myth 10: Axolotls Require Little Upkeep and Are Easy to Take Care of
Disproving:

Although axolotls make interesting pets, it's a fallacy that they require little upkeep and are easy to care for. Axolotls need certain environmental conditions to thrive, such as a balanced food, the right tank configuration, and ideal water quality. Axolotl caretakers must devote time to keeping an eye on water parameters, doing routine maintenance, and furnishing a comfortable environment for their charges.

Myth 11: You Don't Need a Heater for Axolotls

Disproving:

Contrary to popular belief, axolotls are cold-water amphibians, thus they do not require a heater. For axolotls to be healthy, the water temperature must be kept constant and suitable. Even though they prefer lower temperatures, in colder climates or during seasonal changes, adding a heater can assist regulate and prevent swings in the tank's temperature.

Myth 12: Axolotls Are Compulsive Companions

Disproving:

Axolotls are long-term companions. It is reckless to believe the misconception that Axolotls are easily interchangeable or may be thrown away when they are no longer useful. Axolotls are living creatures that demand responsibility and dedication. Adopting an Axolotl entails making the commitment to give them the care they require, attend to their needs, and guarantee their wellbeing for the duration of their lives, which may be more than ten years.

Myth 13: Since axolotls produce oxygen, they don't require a filter

Disproving:

Like other aquatic creatures, axolotls are oxygen-deficient. It's a fallacy that because axolotls create oxygen, they don't require a filter. Although plants can produce oxygen manufacturing, a filter is necessary to keep water clean by taking out waste products like ammonia. For axolotls, using a filter contributes to a stable and healthy environment.

Myth 14: Without water, axolotls can survive

Disproving:

It is untrue that axolotls can live without water; they are aquatic amphibians. They can survive for short periods of time without water, but extended exposure can cause stress, dehydration, and respiratory problems. For axolotls to survive and remain healthy, they need a completely aquatic habitat with the right amount of water depth.

Myth 15: Children Make Excellent Pets with Axolotls

Disproving:

Axolotls can make fascinating pets, but it's important to dispel the misunderstanding that small children should only own them. Axolotls require special care because of their sensitive skin and aquatic lifestyle, which may not be suitable for very young children. Adequate care, adult supervision, and a dedication to addressing the requirements of these exceptional amphibians are all components of responsible ownership.

Final Thoughts: Educated Handling of Axolotls

Dispelling myths about axolotls is essential to disseminating correct information and encouraging responsible ownership. Caretakers can make educated judgments regarding the upkeep, habitat, and general well-being of these fascinating aquatic animals by clearing up myths. Axolotls are special and fulfilling pets, and taking good care of them will increase their longevity, health, and happiness factor for people who enjoy the wonders of amphibian life.

Chapter 10

DIY Projects for Axolotls: Using Creativity to Improve Their Environment

Do-it-yourself (DIY) tasks for the environment of your axolotl can be enjoyable and artistic pursuits. These projects give your aquatic friends environmental enrichment in addition to improving the tank's aesthetic appeal. This extensive book will cover a variety of do-it-yourself projects designed specifically for axolotls, from specialized filtration solutions to tank decorations and skins, all aimed to providing these unusual aquatic animals with a healthy and exciting environment.

Handmade Tank Embellishments

1. Cave Secrets:

- Materials: Slate, PVC pipes, or rocks safe for aquariums.

- Instructions: Arrange the components to form cave-like structures that will give axolotls places to hide. If necessary, use aquarium-safe adhesive to bind components together.

2. Driftwood Hide Hollow:
   - Materials: Aquarium-safe silicone and driftwood.
   - Instructions: Use silicone to join multiple driftwood pieces to form a hollow structure. Make sure there are gaps that allow for simple access and investigation.

3. Silk Décor & Plants:
   - Materials: safe decorations, silk aquarium plants.
   - Instructions: Set up decorations and silk plants to resemble an aquatic scene. Make sure they are firmly attached to avoid unintentional consumption.

- 4. Platforms That Float:

- Materials: Aquarium-safe adhesive and Styrofoam.

- Directions: Cut and glue foam pieces to make platforms that float. These offer places to rest close to the water's surface.

5. Personalized Substrate Designs:

- Materials: colorful aquarium sand and safe aquarium substrate.

- Instructions: Use colored sand to make designs or patterns in the substrate. This makes the tank floor more visually appealing.

DIY Water Quality and Filtration

Changes to Sponge Filter:

- Materials: Airstone, sponge filter already in place.

- Instructions: Fit an airstone to the outflow of the sponge filter. This enhances the oxygenation and circulation of water.

Make Your Own Canister Filter:
- Materials: Aquarium pump, filter medium, and lidded plastic container.
- Instructions: Drill holes in the container to allow water to enter and exit. After adding filter material, attach the aquarium pump.

Plant Filtration in Aquariums:
- Materials: Suction cups and aquatic plants, such as pothos.
- Instructions: Put plant roots in the tank by attaching them to suction cups. By absorbing nitrates, the plants aid in water filtration.

Create Your Own Water Testing Station:

- Materials: Labeled vials and a plastic test tube rack.

- Instructions: For ease of monitoring, arrange the vials used for water testing in a test tube rack. Each vial should have the matching parameter labeled.

Homemade Lighting Upgrades

LED Lighting for Tanks:

- Materials: Waterproof enclosure and LED light strip.

- Instructions: For bright, customizable lighting, affix a weatherproof LED light strip to the sides or lid of the tank.

LED Orbs That Float:

- Materials: fishing line and waterproof LED orbs.

- Instructions: Fasten LED globes at varying points in height using fishing line. This produces an adaptable and appealing lighting effect.

Create Your Own Moonlight Effect:
- Materials: PVC pipe with blue LED light.
- How-to: Insert a blue LED light into a PVC pipe. To create the illusion of a moonlight effect at night, place it over the tank.

Personalized Tank Lid with Integrated Lights:
- Materials: hinges, acrylic sheet, and LED light strips.
- Instructions: Glue LED light strips to an acrylic sheet's bottom. Put in hinges to create a personalized tank cover with integrated lighting.

Homemade Oxygenation and Water Current
Filter Outflow Baffle:

- Materials: Sheets of acrylic or plastic.
- Instructions: Make a baffle to spread out the filter outflow and stop the tank from having strong currents.

Bubble Wall:

- Materials: Suction cups, airstone, and airline tubing.
- Instructions: To create a decorative bubble wall, secure the airstone along the walls of the tank by connecting it to airline tubing.

Personalized Waterfall Function:

- Materials: Slate, tubing, and a small water pump.
- Instructions: To build a tiny waterfall feature, direct water from the pump over a slate or rock structure.

Modifiable Water Jet:

- Supplies: water pump, aquarium valve, and PVC pipe.
- Instructions: To alter the water jet power, add an aquarium valve and connect a PVC pipe to the water pump.

Handmade Nutrition and Enhancement

Floating Ring of Food:

- Materials: Suction cups and airline tubing.
- Instructions: To keep floating food in one place during feeding, make a ring out of airline tubing and suction cups.

Handmade Feeding Tongs:

- Materials: Long tweezers appropriate for aquarium use.
- Instructions: To encourage contact while feeding, provide live or frozen food pieces using long tweezers.

Real-Time Food Dispenser:

- Materials: suction cups and a clear, hole-filled acrylic container.
- Instructions: Fill the container with live food, use suction cups to secure it to the tank, and observe as axolotls use the openings to snatch their meal.

Mats for Foraging:

- Materials: aquarium-safe suction cups and PVC bath mat.
- Instructions: Use suction cups to secure the PVC bath mat to the tank. Then, conceal food underneath the mat so that axolotls can forage for it.

Handmade Thermostat

- Tank Cooling Mechanism:
- Materials: PVC pipe, temperature controller, and computer fan.

- Instructions: Use a temperature controller to turn on a computer fan that is attached to a PVC pipe as the outside temperature rises.

Handmade Chiller:

- Materials: Water pump, tubing, and mini fridge.
- Instructions: To create a homemade cooling system, attach a water pump to the refrigerator and run water through it.

Cover for an insulated tank:

- Materials: reflective material and foam board.
- Instructions: To insulate and control the temperature, make a tank cover out of foam board and reflective material.

Fans for submersible cooling:

- Materials: Cooling fans for submersible aquariums.

- Instructions: To keep the tank at a cool temperature, install submersible cooling fans at key points.

Handmade Themes and Backgrounds

1. 3D Walls:

- Materials: Epoxy suitable for aquariums and Styrofoam.
- Instructions: To create a 3D background, carve styrofoam into the required shapes, apply aquarium-safe epoxy, and attach to the back of the tank.

2. Theme of Naturalism:

- Materials: Live plants, rocks, and driftwood safe for aquariums.
- Instructions: Arrange actual plants, driftwood, and pebbles to create a realistic scene that reflects the natural habitat of axolotls.

3. Ruins or an Underwater Castle:

- Materials: Aquarium-safe glue and ornaments.

- Instructions: For a fanciful theme, use adhesive and safe aquarium ornaments to create an underwater castle or ruins.

4. Personalized Graphic Background:

- Materials: Printed graphics and vinyl sticky sheets.

- Instructions: To create a customized background for the tank, print the required graphics onto vinyl adhesive sheets and stick them on the rear.

Homemade Breeding Configuration

Divider for Breeding Tank:

- Materials: silicone safe for aquariums and acrylic sheet.

- Instructions: To split the breeding couple from the remainder of the tank, make a divider out of silicone and an acrylic sheet.

Egg Gathering Plate:

- Materials: Aquarium-safe adhesive and a smooth dish.

- Instructions: For simple egg collection, use aquarium-safe adhesive to attach a smooth dish to the bottom of the tank.

Platform for Floating Breeding:

- Materials: mesh netting and Styrofoam.

- Instructions: To create a floating breeding platform, cut a piece of styrofoam, cover it with mesh netting, and fasten it to the side of the tank.

Handmade Egg Container:

- Materials: Sponge filter and plastic container.

- Instructions: To make a homemade hatchery for growing axolotl larvae, place a sponge filter inside a plastic container.

Safety Things to Know Before Doing DIY Projects

Resources:

- Make sure your axelotls are healthy by only using products that are safe for aquariums.
- Steer clear of things that could contaminate water with dangerous toxins.

Adhesives:

- Select adhesives safe for aquarium use when joining components together.
- Let adhesives fully dry and properly clean them before introducing Axolotls.

Sharp Points:

- To keep axelotls safe from harm, sand or file down any sharp edges on homemade decorations.

Paints and Dye

- If you intend to color or alter decorations, use only paints or dyes that are safe for aquariums.
- Before putting painted or dyed goods in the tank, be sure they have fully cured.

Tidiness:

- Before adding any DIY things to the tank, give them a thorough cleaning and rinse.
- To keep your surroundings healthy, check and clean DIY projects on a regular basis.

Keep an eye out for behavioral changes:

- After introducing DIY projects, watch for any indications of tension or discomfort in axolotls.

- If a project has a detrimental effect on how Axolotls behave, remove it or change it.

Final Thoughts: Customizing the Habitat of Your Axolotl

Doing do-it-yourself tasks for your axolotl's tank fosters creativity and gives these intriguing aquatic animals a stimulating and enriching environment. With the help of do-it-yourself projects, caregivers can customize the tank to meet the specific needs and tastes of their Axolotls, from unique decorations and hides to creative filtration methods and themed backgrounds. Caretakers can create a customized and aesthetically beautiful environment for their aquatic companions while also improving their well-being and happiness by implementing safety precautions and utilizing materials safe for aquariums.

FAQs

Axolotl: What is it?

A unique aquatic amphibian, the axolotl retains juvenile characteristics throughout its life, a trait known as neotenic features.

Are axolotls in need of a water filter?

A: In order to preserve water quality and eliminate waste, axelotls do indeed benefit from water filters.

Can an axolotl survive without water?

Axolotls are generally aquatic animals and should not be kept out of the water for prolonged periods of time, even if they can withstand brief exposures.

How large do axolotls get?

Axolotls usually grow to a length of 10 to 12 inches, although some can grow as long as 18 inches.

Are axolotls tamed animals?

Axolotls do not chew their little teeth; they are primarily utilized for grasping.

What's for dinner, axolotls?

A: Axolotls consume a range of meals, such as commercial Axolotl pellets, bloodworms, earthworms, and live or frozen brine shrimp.

Can axolotls undergo color changes?

Axolotls can vary in color, albeit they rarely do so dramatically.

When is the right time to feed your axolotl?

A: Feed adult axelotls two to three times a week; for juveniles, feed them less frequently depending on their age and size.

Is it possible to house axolotls together?

A: Although axolotls can live in groups, keep an eye out for aggressive behavior and give them lots of space to relieve stress.

Is a heater necessary for axolotls?

A steady atmosphere is important, and in colder areas, a heater can be required. Axolotls prefer colder temperatures.

Can fish coexist with axolotls?

Axolotl fins and gills can be nibbled by certain fish, thus it's important to choose tank mates carefully.

Do axolotls require UVB light sources?

A: UVB lighting is not necessary for axolotls, however it can be advantageous to their general health and wellbeing.

What is the lifespan of axolotls?

Axolotls in captivity can survive for ten to fifteen years if given the right care.

Can internal organs be regenerate in axolotls?

Axolotls have the ability to regenerate several internal organs, albeit the degree of regeneration varies.

How frequently should the Axolotl tank be cleaned?

A: To preserve water quality, do weekly partial water changes and comprehensive tank cleaning every few weeks.

Can you manage axolotls?

A: To prevent stress and potential harm to their sensitive skin, handle axolotls lightly and with damp hands.

Do axolotls produce noises?

Axolotls don't usually talk, but while they're courting, they might make little clicking noises.

Can you keep an axolotl in tap water?

A water conditioner should be used to eliminate chlorine and chloramine from tap water before using it.

How are axolotls able to breathe?

Axolotls are able to take in oxygen from the air and water because they breathe through both their lungs and gills.

Are axolotls able to climb out of a tank?

Axolotls are not good climbers, however to avoid unintentional escapes, secure tank lids are advised.

Do axolotls require a particular kind of substrate?

A: To avoid ingestion and lower the chance of impaction, bare earth or fine sand are advised.

Are axolotls able to survive in brackish water?

Axolotls are amphibians native to freshwater, hence they shouldn't be housed in brackish water.

Axolotls require a water current, right?

Axolotls love to swim in still or mild water; strong currents should be avoided.

Axolotl sex: How can I do that?

A: During breeding season, males may have a more pronounced cloaca and apparent swelling, but sexing can be difficult.

Can an axolotl live by itself?

A: Because axolotls are solitary creatures, keeping them by themselves may lessen stress and hostility.

Do axolotls have a home in ponds?

Axolotls can live in outdoor ponds, but it's important to keep in mind things like temperature changes and possible predators.

Axolotls: Do they hibernate?

Axolotls don't hibernate, however in colder climates, they could become less active.

Is axolotl vision impaired in the dark?

Axolotls can see just a limited amount in low light, but they are used to it.

Is there a sleep routine for axolotls?

Axolotls are crepuscular animals, meaning that dawn and dusk are when they are most active.

Do axolotls have access to feeder fish?

A: It's not advisable to feed feeder fish to an axelotl because they can harbor parasites or diseases.

Can you train an axolotl?

Axolotls may come to link specific behaviors with eating, but they do not react to teaching in the conventional sense.

Is it possible to house axolotls with various varying sizes?

A: To avoid possible violence, do not keep significantly larger Axolotls alongside smaller ones.

Do axolotls molt their hides?

A: Axolotls occasionally lose their skin, which might take the form of a white film. Make sure your tank is clean when you shed.

Is it possible to house axolotls in a tiny tank or bowl?

Axolotls need large tanks; their wellbeing is not enhanced by keeping them in tiny bowls.

Can an axolotl leap from a tank?

A: Even while axolotls don't typically jump, it's still a good idea to have a tight-fitting tank cover to stop any unintentional escapes.

Can axolotls and turtles coexist in a home?

A: Stay away from housing axolotls alongside turtles because the latter may bite the axolotls' fragile gills.

Do axolotls harbor infections that are dangerous to people?

Axolotls are rare carriers of Salmonella, thus handling them requires extreme caution and good hygiene.

Can axolotls repair damage to their gills?

Axolotls can partially regenerate their damaged gill tissue, but wounds should be prevented.

Is it possible to keep axolotls alongside frogs?

A: It's ideal to keep axolotls and frogs in separate housing because they require distinct types of care.

Is it possible to keep axolotls on a gravel substrate?

A bare bottom or fine sand is a safer option for the tank because gravel presents a choking hazard.

Can axolotls only consume pellets?

A diversified diet containing live or frozen meals is advised for healthy nutrition, even though commercial pellets are handy.

Can an axolotl grow new eyes?

Axolotls have the ability to regenerate certain eye tissues, but not the full eye.

Do axolotls know who their owners are?

Axolotls may grow acclimated to the presence of their caregiver, but unlike mammals, they do not show signs of recognition.

Is it possible to keep axolotls in a community tank?

A: Because they might see their smaller tank mates as possible prey, axolotls are best kept in species-exclusive tanks.

Can axolotls survive by drinking tap water alone?

A water conditioner should be used to eliminate any dangerous contaminants from tap water before using it.

Can an axolotl colonize a bowl fitted with a filter?

A: Even with a filter, axolotls require a larger space for their wellbeing than a bowl.

Can you eat insects with axolotls?

Axolotls have the ability to eat little, harmless insects.

Is it possible to keep axolotls with shrimp?

Axolotls may consider shrimp to be prey, and their presence may cause problems.

Can you feed axolotls vegetables?

Axolotls should not be fed vegetables as their main source of nutrition because they are predominantly carnivorous, however they may nibble on them.

Can an axolotl experience stress?

A: Absolutely, axolotls can become stressed. To reduce stress, caregivers can create a peaceful and comfortable atmosphere.